The Facts on Matter

What are COMPOUNDS?

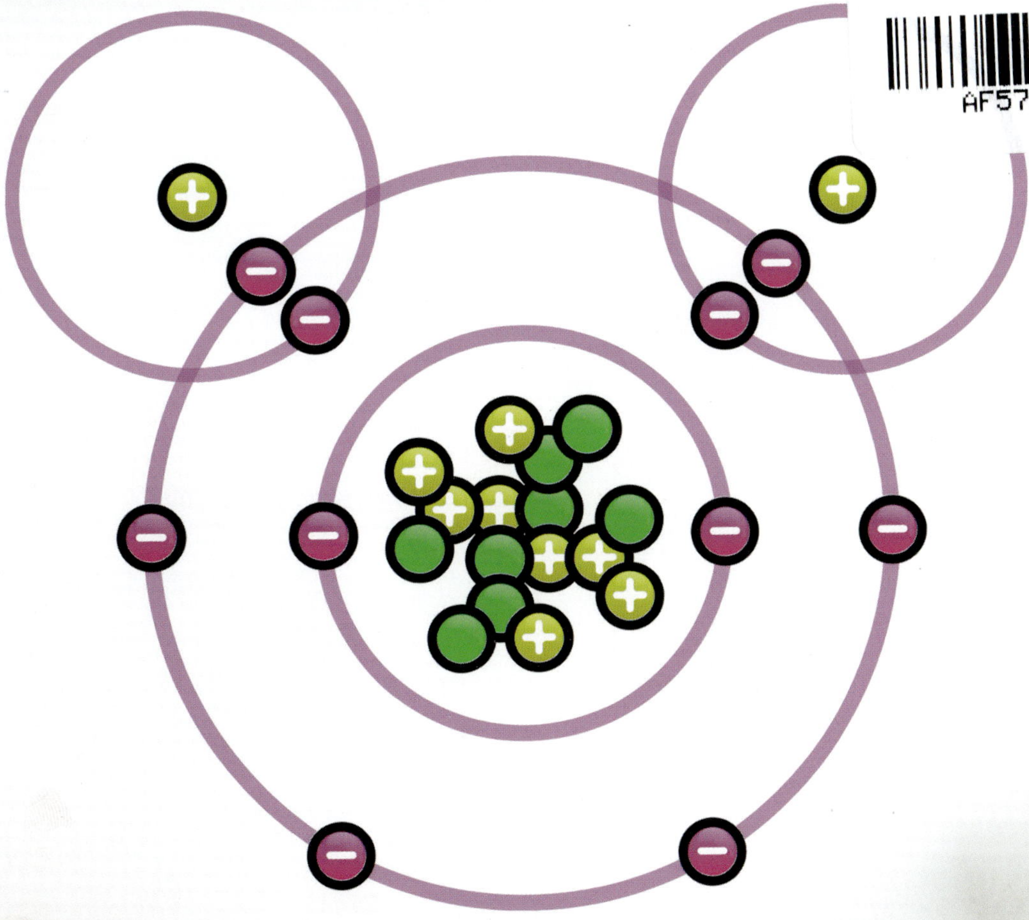

By Elise Tobler

Please visit our website, www.garethstevens.com. For a free color catalog of all our high-quality books, call toll free 1-800-542-2595 or fax 1-877-542-2596.

Library of Congress Cataloging-in-Publication Data

Names: Tobler, Elise, 1970- author.
Title: What are compounds? / Elise Tobler.
Description: New York : Gareth Stevens Publishing, 2022. | Series: The facts on matter | Includes index.
Identifiers: LCCN 2020033189 (print) | LCCN 2020033190 (ebook) | ISBN 9781538267059 (library binding) | ISBN 9781538267035 (paperback) | ISBN 9781538267042 (set) | ISBN 9781538267066 (ebook)
Subjects: LCSH: Chemical elements–Juvenile literature. | Inorganic compounds–Juvenile literature. | Organic compounds–Juvenile literature. | Chemistry–Juvenile literature.
Classification: LCC QD466 .T72 2022 (print) | LCC QD466 (ebook) | DDC 546./2–dc23
LC record available at https://lccn.loc.gov/2020033189
LC ebook record available at https://lccn.loc.gov/2020033190

First Edition

Published in 2022 by
Gareth Stevens Publishing
29 E. 21st Street
New York, NY 10010

Designer: Katelyn E. Reynolds
Editor: Char Light

Photo credits: Cover, pp. 1 (compound), 7, 9, 13, 17 VectorMine/ iStock / Getty Images Plus; cover, pp. 1–24 (background) sudanmas/E+/Getty Images; cover, pp. 1–24 (chemistry doodles) backUp/Shutterstock.com; cover, pp. 1–24 (banner) Ozz Design/Shutterstock.com; cover, pp. 1–24 (paper) Sergey Mironov /Shutterstock.com; p. 5 Burke/Triolo Productions / Photolibrary / Getty Images Plus; p. 7 VGstockstudio/Shutterstock.com; p. 9 Kristin Duvall/ The Image Bank / Getty Images Plus; p. 11 Martin Leigh/ Oxford Scientific /Getty Images Plus; p. 13 joshblake/E+/Getty Images; p. 19 Tim Wright/Corbis Documentary/Getty Images; p. 21 SolStock/E+/Getty Images.

Printed in the United States of America

CPSIA compliance information: Batch #CWGS22: For further information contact Gareth Stevens, New York, New York at 1-800-542-2595.

Find us on  

CONTENTS

What Is a Compound? 4
Properties of Compounds 6
Solid Compounds 8
Liquid Compounds 10
Gas Compounds 12
Compounds vs. Mixtures 14
Chemical Bonds 16
Organic and Inorganic Compounds 18
Compounds In Your World 20
Glossary 22
For More Information 23
Index 24

Words in the glossary appear in **bold** type the first time they are used in the text.

What Is a COMPOUND?

Everything is made of **atoms**, including your body and everything you own. Elements, like the oxygen you breathe, are made of only one kind of atom. When atoms bond, or join together, they create **molecules**.

You can combine elements to create entirely new things, kind of like how you can combine some words to make a new word ("steam" and "boat" can become "steamboat"). Compounds are the combination of two or more different elements into one new molecule.

Know the Basics!

Compound names can tell you how many atoms are inside a compound. We know "sodium trichloride" has one atom of sodium and three atoms of chloride because "tri" means "three."

Water is most likely the compound you know best! It's a combination of the elements hydrogen and oxygen.

Properties of COMPOUNDS

Compounds are like **recipes** you might use in the kitchen. You combine single elements to make something new. The "recipe" for water is always two parts hydrogen and one part oxygen. Every combination of elements will make an entirely new compound. The combinations are nearly endless!

Compounds have **properties** different from the elements that combine to make them. Water is often a liquid, but the individual elements of hydrogen and oxygen are usually gases.

Know the Basics!

Science is the study of the world. Chemistry is one type of science. It studies the properties of matter and how matter works with **energy**.

Cooking uses a lot of chemistry. These eggs, flour, and butter are combining to make batter!

Solid COMPOUNDS

In science, when we say something is solid, we mean that it's hard to the touch and keeps its shape even when it isn't in a container. Solid compounds are **unique** because they take longer to make than liquid or gas compounds.

The salt in your kitchen is a solid compound. It's made of the elements sodium and chlorine. The "recipe" is one part sodium and one part chlorine to become "sodium chloride."

Know the Basics!

Matter comes in four states: solid, liquid, gas, and **plasma**. Compounds can be made in any state of matter.

Salt is held together by ionic bonds. In ionic bonds, one element gives **electrons** to the other. In this case, sodium gives chlorine an electron.

Liquid COMPOUNDS

A liquid, having no shape of its own, shapes itself to whatever container it may be in. Outside of a container, liquid will flow to the lowest point it can find.

Touching water and touching oil feels very different to us because these compounds have different properties. But they're still both compounds, and they're both often found as liquids! One molecule of water only has three atoms in it. Cooking oils are made up of many more atoms of the elements carbon, hydrogen, and oxygen.

Just because two **substances** are mixed together, doesn't mean they will form bonds. Oil and water don't!

Gas COMPOUNDS

When you add energy to matter or take energy away from matter, you cause it to change its state. When you heat water, you make steam, which is the gas form of water. Compounds may also be found when matter is in this state.

You may know the gas propane because it's used with outdoor grills. Propane is three parts carbon and eight parts hydrogen. Another gas compound you might know is carbon dioxide. This is the gas humans and many animals breathe out!

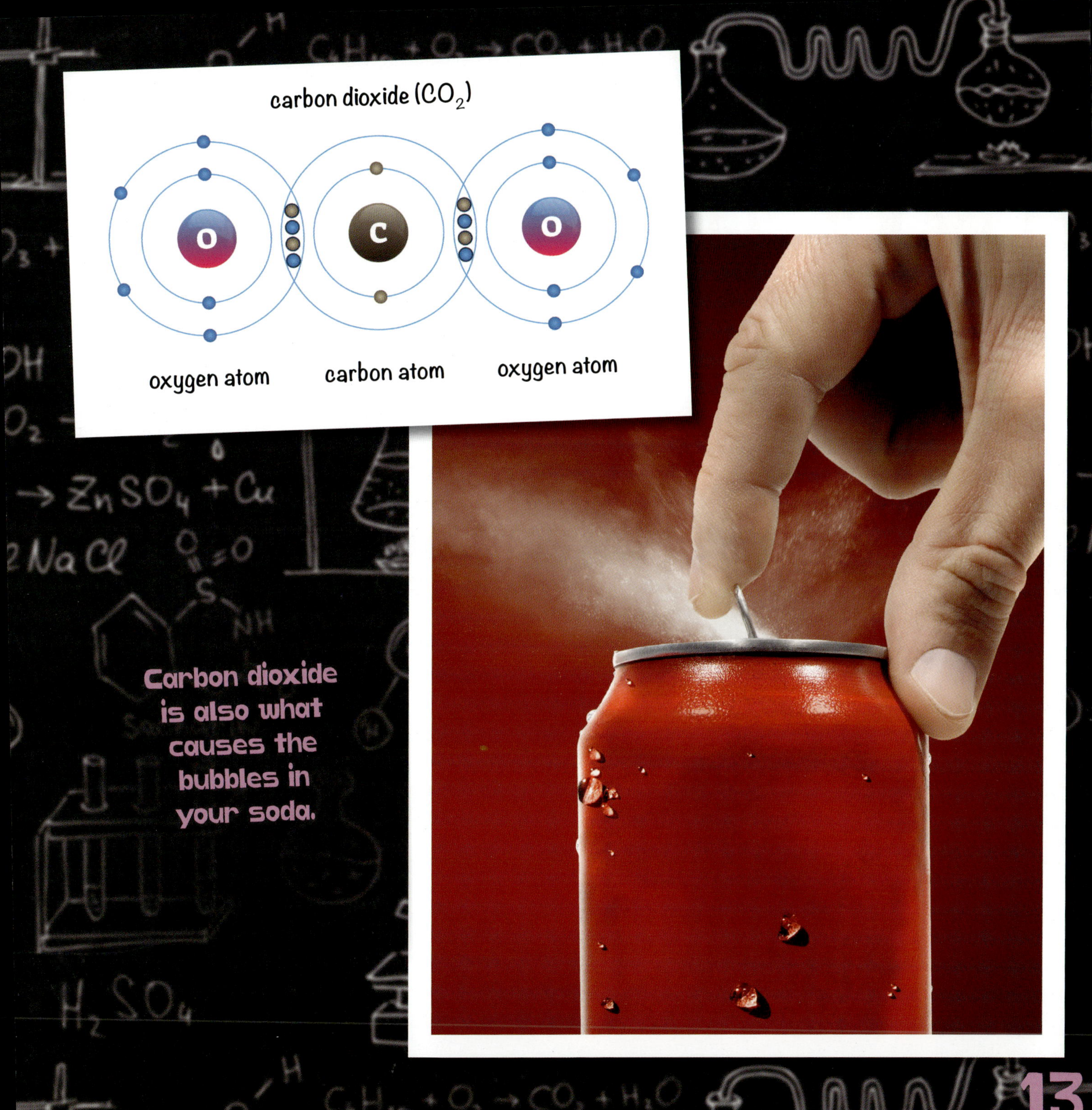

Carbon dioxide is also what causes the bubbles in your soda.

Compounds vs. MIXTURES

A mixture is the combining of two or more substances that doesn't include any **chemical reaction**. A mixture's atoms stay separate. Mixtures can be taken apart by using properties of the different atoms.

The atoms in compounds join in a chemical reaction. This makes it much harder to take compounds apart.

While the salt in your kitchen is a compound, a salad is a mixture. The elements in a grain of salt cannot be easily separated, but you can take pieces out of your salad.

IS IT A MIXTURE OR A COMPOUND?

Questions	Compound	Mixture
What's the "recipe" like?	Each compound has its own "recipe" that must be the same every time.	You can add more or less of each item.
How does it form?	A chemical reaction bonds the atoms.	Items mix together but do not join chemically.
What are its properties?	The compound has different properties from the single elements that combine to make it.	Each item in the mixture keeps its own properties.
Can it be separated?	Items can only be separated by a chemical reaction.	Each item is easy to separate from the mixture.
Examples	water, carbon dioxide, sodium chloride	salad, mud (dirt and water)

Compounds and mixtures each have different properties.

Chemical BONDS

Compounds need a chemical reaction to form. Every atom has a shell surrounding it, much like an egg. Atomic shells like to be full. If an atom has a gap in its shell, it fills the gap by getting close to another atom.

When atoms share electrons, it's called a covalent bond. When one atom gives another atom an electron so they can both have full shells, it's an ionic bond. Ionic bonds are stronger than covalent bonds. These bonds are why it isn't easy to separate compounds.

Know the Basics!

An ion is an atom has extra or missing electrons.

IONIC BOND

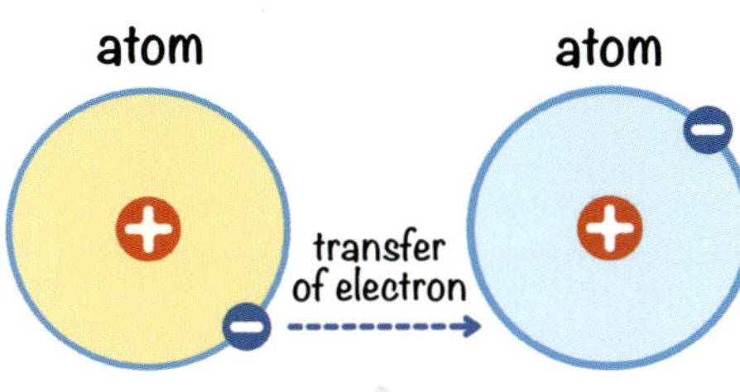

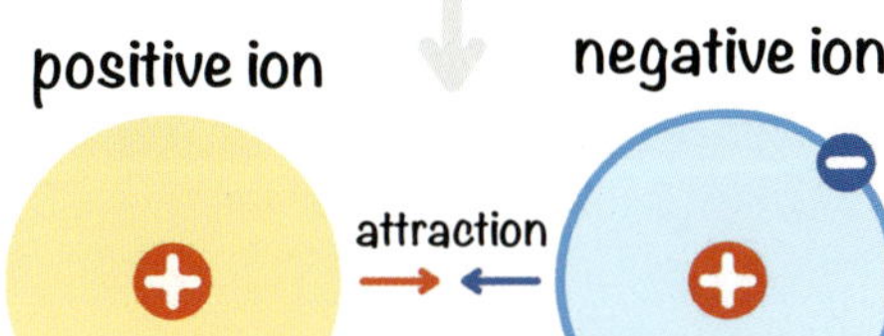

ionic bond

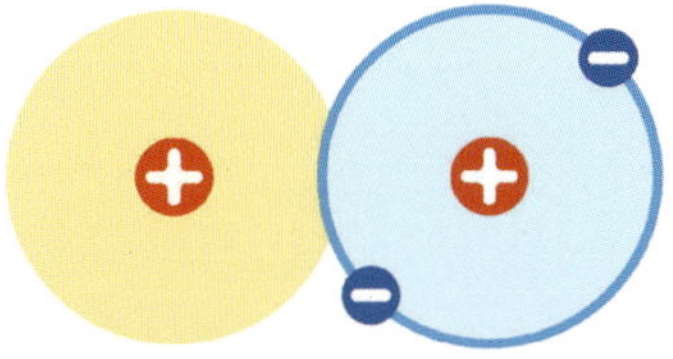

COVALENT BOND

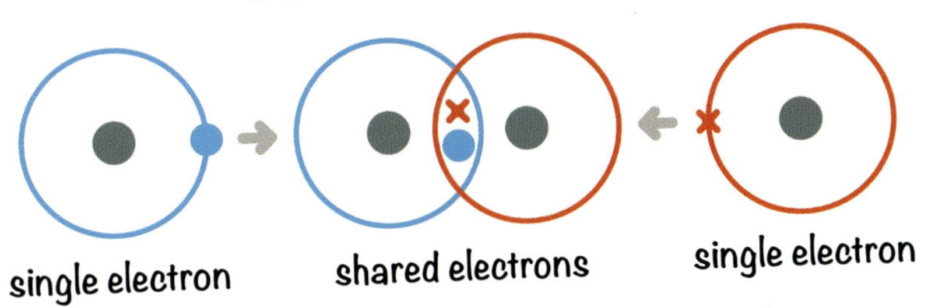

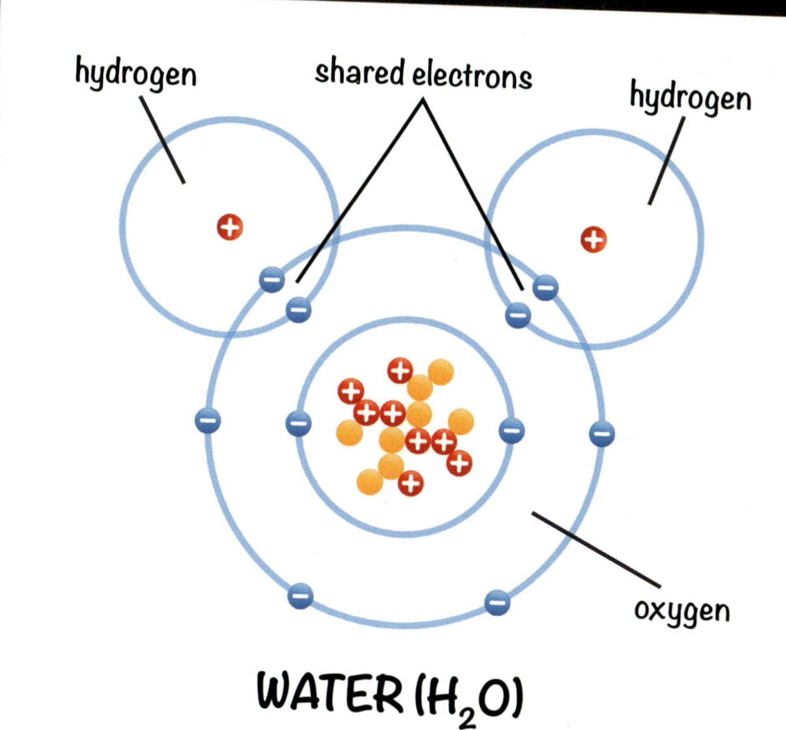

WATER (H_2O)

Hydrogen and oxygen have a covalent bond when they combine to make water.

Organic and Inorganic COMPOUNDS

Organic compounds are found in all living things. In fact, the word "organic" means having to do with living things! These compounds always contain carbon. They make up the food you eat and your body itself.

Inorganic means something made of non-living matter. Inorganic compounds are those that don't contain both hydrogen and carbon. Many do contain just hydrogen though. Most of Earth's crust is made of inorganic compounds. Inorganic compounds include metals and salts.

Charcoal is largely carbon.
It's often made from wood,
making it an organic compound.

Compounds In Your WORLD

Today, doctors are using organic compounds to help treat illnesses. The medicines aspirin and penicillin are both compounds that help people live longer, healthier lives.

The world has more compounds than you can count! As we use science to study the world around us, we will most likely discover even more compounds.

Know the Basics!

Synthetic compounds, such as plastic, may be made of natural parts, but do not occur in nature.

Some clothing, like a jacket, is made of synthetic compounds.

GLOSSARY

atom: one of the smallest bits of matter

chemical reaction: the process that turns one substance into another

electron: a particle found in atoms that acts as a carrier of electricity in solids

energy: power used to do work

molecule: a very small piece of matter made up of atoms bonded together

plasma: a hot gas made up of ions and electrons

property: a special quality or characteristic of something

recipe: an explanation of how to make food

substance: a certain kind of matter

synthetic: having to so with something made by combining different substances

unique: one of a kind

For More INFORMATION

Books

Gray, Theodore. *Molecules: The Elements and the Architecture of Everything.* New York, NY: Black Dog & Leventhal, 2018.

Heinecke, Liz Lee. *The Kitchen Pantry Scientist: Chemistry for Kids: Homemade Science Experiments and Activities Inspired by Awesome Chemists, Past and Present.* Beverly, MA: Quarry Books, 2020.

Slingerland, Janet. *Explore Atoms and Molecules!: With 25 Great Projects.* White River Junction, VT: Nomad Press, 2017.

Websites

Chemical Compound Facts For Kids
kids.kiddle.co/Chemical_compound
Compounds are made easy at this site, with links for in-depth exploration.

Chemistry For Kids
www.ducksters.com/science/chemistry/
Explore everything about chemistry at this site, which includes easy at-home experiments.

Compound Basics
www.chem4kids.com/files/atom_compounds.html
A great place to begin learning about compounds.

INDEX

atomic shell 16

chemistry 6, 7

covalent bond 16, 17

element 4, 5, 6, 8, 9, 10, 14, 15

gas 6, 8, 12

inorganic 18

ion 16, 17

ionic bond 9, 16, 17

liquid 6, 8, 10

matter 6, 8, 12, 18

naming compounds 4

organic 18, 19, 20

science 6, 8, 20

solid 8